BLOOD TRANSFUSION TRANSMITTED INFECTIONS – A REVIEW

Dr. Gunvanti Rathod

Dr. Pragnesh Parmar

Dr. Sangita Rathod

Dr. Ashish Parikh

BLOOD TRANSFUSION TRANSMITTED INFECTIONS – A REVIEW

Dr. Gunvanti Rathod, MD (Pathology)

Assistant Professor
Department of Pathology
SBKS Medical Institute and Research Centre
Vadodara, Gujarat, India

Dr. Pragnesh Parmar, MD (Forensic Medicine)

Assistant Professor
Department of Forensic Medicine
SBKS Medical Institute and Research Centre
Vadodara, Gujarat, India

Dr. Sangita Rathod, MD (Medicine)

Assistant Professor
Department of Medicine
AMCMET Medical College
Ahmedabad, Gujarat, India

Dr. Ashish Parikh, MD (Medicine)

Consultant Physician
Gayatri Hospital
Gandhinagar, Gujarat, India

DEDICATION

This book is dedicated to my loving daughter **Jayani**.

-Dr. Gunvanti Rathod

ACKNOWLEDGEMENTS

We acknowledge the immense help received from the scholars whose articles are cited and included in references of this book. The authors are also grateful to authors / editors / publishers of all those articles, journals and books from where the literature for this book has been reviewed and discussed.

We express our gratitude to our parents and in-laws for their constant encouragement, support and blessings.

It will be an injustice if we do not thank all our students for their innovative ideas and feedback.

CONTENTS

INTRODUCTION

- Timely transfusion of blood saves millions of lives, but unsafe transfusion practices puts millions of people at risk of transfusion transmissible infections (TTI's). [1] An unsafe blood transfusion is very costly from both human and economic points of view. Morbidity and mortality resulting from the transfusion of infected blood have far-reaching consequences, not only for the recipients themselves, but also for their families, their communities and the wider society.

- It should, therefore, be obligatory on those who are involved in transfusion of blood to a patient for saving his life, that the blood transfusion does no harm to the patient. Nothing could be worse than the fact that in an attempt to save life, blood & blood products having transmissible infectious agents have been given. Many of these infectious agents may cause death or prolonged illness. Hence it is necessary to understand the organisms which could be transmitted through blood transfusion and means by which this could be prevented.

- With the emergence of HIV/AIDS there is great demand for guaranteed safe blood and blood products. Today there is rigorous screening of blood to prevent transmission of blood borne infection, the transfusion of blood products have already achieved high level of safety. The "window period" viraemia can be further reduced by screening the donated blood for nucleic acid testing methods. Thus further reducing the risk and increasing the already high cost of testing.

- The goal of a safe and affordable blood supply that can meet the growing global demands may be reached by the coordinated optimization of each step in the transfusion chain, including the careful consideration of donor eligibility criteria, adherence to rigorous rules during donation, processing and storage, the optimal implementation of available screening tests, the use of suitable pathogen inactivation methods and finally the vigilance of prudent physicians, who evaluate the necessity of each transfusion. [2]

•Since a person can transmit an infection during its asymptomatic phase, transfusions can contribute to an ever widening pool of infection in the population. The economic costs of the failure to control the transmission of infection include increased requirement for medical care, higher levels of dependency and the loss of productive labor force, placing heavy burdens on already overstretched health and social services and on the national economy.

CRITERIA FOR SELECTION OF BLOOD DONORS [3]

•Accept only voluntary/replacement non-remunerated blood donors if following criteria are fulfilled.

1. The interval between blood donations should be no less than three months.

2. The donor shall be in good health, mentally alert and physically fit and shall not be a jail inmate or a person having multiple sex partners or a drug-addict.

3. The donor shall be in the age group of 18 to 60 years.

4. The donor shall not be less than 45 kilograms.

5. Temperature and pulse of the donor shall be normal (Pulse - between 50 and 100/minute with no irregularities, oral temperature not exceeding 37.5 C).

6. The systolic and diastolic blood pressures are within normal limits without medication (Systolic 100-180 mm Hg and Diastolic 50 - 100 mm Hg)

7. Haemoglobin shall not be less than 12.5 g/Dl.

8. T

he donor shall be free from acute respiratory diseases.

9.   T

he donor shall be free from any skin disease at the site of phlebotomy.

10.   T

he donor shall be free from any disease transmissible by blood transfusion, in so far as can be determined by history and examination indicated above.

11.   T

he arms and forearms of the donor shall be fee from skin punctures or scars indicative of professional blood donors or addiction of self-injected narcotics.

• **D**

efer the donor permanently if suffering from any of the following diseases:

1.   C
   ancer
2.   H
   eart disease
3.   A
   bnormal bleeding tendencies
4.   U
   nexplained weight loss
5.   D
   iabetes
6.   H
   epatitis B infection
7.   C
   hronic nephritis
8.   Si
   gns and symptoms, suggestive of AIDS
9.   Li
   ver disease
10.   T
    uberculosis
11.   P
    olycythemia Vera
12.   A
    sthma
13.   E
    pilepsy
14.   L
    eprosy
15.   S
    chizophrenia
16.   E
    ndocrine disorders
17.   It
    is important to ask donors if they have been engaged in any risk behavior.

- Allow sufficient time for discussion in the private cubicle. Try and identify result- seeking donors and refer them to VCTC (Voluntary Counseling and Testing Center). Reassure the donor that strict confidentially is maintained.

Who is Eligible to Donate Blood? [79]

- Any healthy adult, both male and female, can donate blood. Men can donate safely once in every three months while women can donate every four months.

EXCLUSION CRITERIA:

1. Any donor not meeting all criteria's for eligibility of blood donation
2. Any eligible donor having any kind of reaction during the blood donation procedure will be excluded from the studies
3. Any defects found in the sample collected (Bag Leakage, Improper Maintenance Of Cold Chain During Transportation, Preservation Defects, Temperature Defects, Any Undesirable Physical And Biochemical Changes In Stored Blood)

HISTORICAL BACKGROUND

• Blood has always held mysterious fascination for all and is considered to be the living force of our body. Ancient Egyptians recognized the life giving properties and they used it for baths to resuscitate the sick, rejuvenate the old and infirm, as a tonic for treatment of various disorders. [9]

• The year 1942 it often credited with first human transfusion that was given to **Pope Innocent VII** who had suffered a stroke previous year. [10] In the early 1600, **William Harvey** discovered the circulatory system in the human body, while the work of **Wren, Wilkins, Boyle** and **Willis** demonstrated that intravenous injection of substances was possible. Both of these discoveries opened the way for modern transfusion medicine. [11]

• The idea of transfusing the blood was first described in 1615 by **Andreas Libavius**, a chemist and physician in Saxony. He wrote a procedure using Male/Female silver tubes to connect the arteries of a young man to those of an old man. He believed that hot and spirituous blood of young man will pour into the old one as it were from a fountain of life and all his weakness will be dispelled. [12]

• First successful transfusion, performed by **Richard Lower** at Oxford in February 1665, was published on November 19, 1666, in the Philosophical Transactions of the Royal Society Transactions (Transactions) in a short notation titled, "The Success of the Experiment of Transfusing the Blood of One Animal into Another." [13]

• The first human transfusion was then performed on June 15, 1667, when **Denis** administered the blood of a lamb to 15 years old boy. Three ounces of the boy's blood were exchanged for 9 ounces of lamb arterial blood. [14]

• Although the first two subjects who underwent transfusion by **Denis** were not adversely affected, the third and fourth recipients died. The death of the third subject was easily attributable to other causes. However, the fourth case initiated a sequence of events that put an end to transfusion for 150 years. **Anthony du Mauroy** was a 34 years old man who suffered from

intermittent bouts of maniacal behavior. On December 19, 1667, **Denis** and his assistant **Paul Emmerez** removed 10 ounces of the man's blood and replaced it with 5 or 6 ounces of blood from the femoral artery of a calf. After the second transfusion, **du Mauroy** experienced a classic transfusion reaction. He awoke the following morning and "made a great glass full of urine, of a colour as black, as if it had been mixed with the soot of chimneys". The physicians of Paris strongly disapproved the experiments in transfusion. Three of them approached du Mauroy's widow and encouraged her to lodge a malpractice complaint against Denis. [15] The court also stipulated "that for the future no Transfusion should be made upon any Human Body but by the approbation of the Physicians of the Parisian Faculty." [16] At this point, transfusion research went into decline, and within 10 years it was prohibited in both France and England.

• After the edict that ended transfusion in the 17th century, the technique laid dormant for 150 years. Stimulated by earlier experiments by **Leacock**, transfusion was "resuscitated" and placed on a rational basis by James Blundell (1790-1877), a London obstetrician. [17]

•The frequency of postpartum hemorrhage and death troubled **Blundell**. In 1818 he wrote: "**A few months ago I was requested to visit a woman who was sinking under uterine hemorrhage . . . Her fate was decided, and not withstanding every exertion of the medical attendants, she died in the course of two hours.**" [18] Because of his extensive experimental transfusion studies in both animals and humans, Blundell is widely accepted as "Father **of Blood Transfusion**". [13]

•In 1869, **Braxton-Hicks**, using blood anti-coagulated with phosphate solutions, performed a number of transfusions on women with obstetric bleeding. Many of the patients were in extremis and ultimately all died. Frustration with blood as a transfusion product led to even more bizarre innovations. From 1873 to 1880, cow, goat, and even human milk were transfused as a blood substitute. [19] Sodium citrate was widely considered to be the most effective and non-toxic anticoagulant. Various scientists, **Levisohn** of New York (established the optimal citrate concentration for

anticoagulation) [20], **Hustin** of Brussels and **Weil** of New York (demonstrated the feasibility of refrigerated storage) claimed credit the first to be use in human blood transfusion.

•The 20th century was ushered in by a truly monumental discovery. In 1900, **Karl Landsteiner** (1868-1943) observed that the sera of some persons agglutinated the red blood cells of others. With the identification of **blood groups A, B, and C** (subsequently renamed group O) by Landsteiner and of **Group AB** by **Decastello** and **Sturli**, the stage was set for the performance of safe transfusion. [21] For this work, Landsteiner somewhat belatedly received the Nobel Prize in 1930. **Rous** and **Turner** developed the **anticoagulant** solution that was used during World War I. [22]

•**Fantus** was the first to coin the phrase **"Blood Bank"** for the operation because blood could be stored and saved for future use. **Bernard Fantus** as the Director and **Oswald Robertson** as an advisor established the first blood bank at Cook County Hospital in Chicago in 1937. [23] In this latter facility, blood was collected into glass flasks containing sodium citrate, sealed, and stored refrigerated. Pilot tubes were prepared for typing and serology testing.

•In early 1940, the anticoagulant preservative Acid-Citrate-Dextrose (ACD) was developed in great Britain by **Loutit** and **Mollison** which extended the shelf life of whole blood to 21 days, could be autoclaved, and had the advantage of being easier to prepare, while requiring a smaller volume of solution relative to the amount of blood. [24] **Gibson** group in 1950 used Citrate phosphate dextrose (CPD) solution as blood could be stored for up to 28 days with better red cell survival than ACD. [25]

•By 1960s, preservative solution containing Adenine were shown to greatly extend shelf life of stored refrigerated blood as compared to ACD or CPD alone. The Food and Drug Administration in 1978 approved the addition of Adenine to CPD to create CPDA-1, which increased the shelf life of blood to 35 days.

•By 1983, the Food and Drug Administration (FDA) approved additive solutions containing saline, adenine and dextrose for RBC's , extending shelf life of this component to 42 days. [26]

HUMAN IMMUNODEFICIENCY VIRUS (HIV)

World History

•**1981** - AIDS was first reported.

•**1982** - FDA received first IND submission for treatment of AIDS.

•   **1984** - AIDS identified as being caused by a human retrovirus, Human Immunodeficiency Virus (HIV).

•   **1985** - FDA approved first enzyme linked immunosorbent assay (ELISA) test kit to screen for antibodies to HIV.

- **1987** - On March 19, FDA approved AZT - the first drug approved for the treatment of AIDS.

  - On April 29, FDA approved the first Western blot blood test kit - a more specific test. [49]
  - On August 18, FDA sanctioned the first human testing of a candidate vaccine against HIV.

- FDA Published regulations which require screening all blood and plasma collected in the U.S. for HIV antibodies.

- FDA completed studies demonstrating the safety of immune globulin products.

- FDA revised its strategy for the regulation of condoms by strengthening its inspection of condom manufacturers and repackers and testing of domestic and imported condoms in commercial distribution, and providing guidance on labeling of condoms for the prevention of AIDS.

Historical Aspect of India [27]

- India's first cases of HIV were diagnosed among sex workers in Chennai, Tamil Nadu. Most of the initial cases had occurred through heterosexual sex; but at the end of the 1980s, a rapid spread of HIV was observed among injecting drug users in Manipur, Mizoram and Nagaland. In 1987, a National AIDS Control Program was launched to coordinate national responses. Its activities covered surveillance, blood screening and health education. In 1992, the government set up NACO (National AIDS Control Organization), to oversee the formulation of policies, prevention work and control programs related to HIV and AIDS. In 2001, the government adopted the National AIDS Prevention and Control Policy. NACP III was launched formally on 6th July 2007.

Magnitude of HIV/AIDS in India [29]

- Based on HIV Sentinel Surveillance 2008-09, it is estimated that India has an adult prevalence of 0.31 % with 23.9 lakh people infected with HIV, of which, 39% are females and 3.5% are children . The estimate highlights an overall reduction in adult HIV prevalence, HIV incidence (new Infections) as well as AIDS related mortality in India. HIV infection has declined by more than 50% during last decade (in 2009 – 1.2 Lakh new HIV infection against 2.7 lakh in 2000)

Etiology of HIV-AIDS [32]

- Human immunodeficiency virus is an RNA virus and belongs to the genus lentivirus (lenti – slow) within the family Retroviridae (retro – backwards), so called because viruses (including HIV) in the family possess a reverse transcriptase (RT) enzyme to convert the viral RNA template into DNA, which integrates in the cellular DNA to cause persistent infection. The other virus in the genus lentivirus is simian immunodeficiency virus (SIV), which infects monkeys.

- There are two known HIV viruses that cause human infection namely HIV 1 and HIV 2. Human immunodeficiency virus 1 is further divided into three groups: 'major' group, M; 'outlier' group, O; and 'new' group, N. Group M has several subtypes or clades (subtypes A to K). Other human viruses in the family Retroviridae are human T-cell leukemia virus (HTLV) HTLV 1 and HTLV 2.

Important Properties of Lentivirus (Nononcogenic Retroviruses) [33]

- **Virion:** Spherical, 80–100 nm in diameter, cylindrical core

- **Genome:** Single-stranded RNA, linear, positive-sense, 9–10 kb, diploid; genome more complex than that of oncogenic retroviruses, contains up to six additional replication genes

- **Proteins:** Envelope glycoprotein undergoes antigenic variation; reverse transcriptase enzyme contained inside virions; protease required for production of infectious virus

- **Envelope:** Present

- **Replication:** Reverse transcriptase makes DNA copy from genomic RNA; provirus DNA is template for viral RNA. Genetic variability is common.

•Maturation:

> Particles bud from plasma membrane
> Outstanding characteristics: Members are non-oncogenic and may be cytocidal
> Infect cells of the immune system
> Proviruses remain permanently associated with cells
> Viral expression is restricted in some cells in vivo
> Cause slowly progressive, chronic diseases
> Replication is usually species-specific
> Group includes the causative agents of AIDS

• **The Genome:** HIV is a retrovirus, a member of the Lentivirus genus, and exhibits many of the physicochemical features typical of the family. The unique morphologic characteristic of HIV is a cylindrical nucleoid in the mature virion.

Structural Genes:

•**gag** – Encodes proteins that form core of virions p17 matrix protein, P24 capsid protein p15 – nucleocapsid precursor processed to p2, p7, p1 and p6

• **pol** – Encodes the enzyme , RT – reverse transcriptase , PR – protease, and IN - integrase.

•**env** – Encodes the envelope glycoproteins gp 120 and gp41.

p to six additional genes regulate viral expression and are important in disease pathogenesis in vivo. Although these auxiliary genes show little sequence homology among lentiviruses, their functions are conserved. (The feline and ungulate viruses encode fewer accessory genes.)

•**tat** - Early-phase replication protein, functions in "transactivation," whereby a viral gene product is involved in transcriptional activation of other viral genes. Transactivation by HIV is highly efficient and may contribute to the virulent nature of HIV infections.

•**rev** - Protein is required for the expression of viral structural proteins. It facilitates the export of unspliced viral transcripts from the nucleus; structural proteins are translated from unspliced mRNAs during the late phase of viral replication.

•**nef** - Protein increases viral infectivity, facilitates activation of resting T cells, and down regulates expression of CD4 and MHC class I. It is necessary for simian immunodeficiency virus (SIV) to be pathogenic in monkeys.

•**vpr** protein increases transport of the viral preintegration complex into the nucleus and also arrests cells in the G2 phase of the cell cycle.

•**vpu** protein promotes CD4 degradation.

•HIV-2 is morphologically similar to HIV-1, and the immunodeficiency state associated with HIV-2 infection is indistinguishable from AIDS caused by HIV-1. The risk factors for infection in both diseases seem to be similar. However, HIV-2 is felt to be derived from a different nonhuman primate virus from HIV-1, and is more difficult to transmit than HIV-1. People infected with HIV-2 tend to progress to AIDS more slowly than those infected with HIV-1.34   Accessory genes carried by HIV include tat, rev, nef, vif, vpr, and vpu (for HIV-1) or vpx (for HIV-2).

Types and Sub Types of HIV

- There are three major groups of HIV-1, based upon phylogenetic analysis, which likely arose from different transmission events in history among primates and humans. These groups are defined as M (major), N (nonmajor and nonoutlier), and O (outlier). Within these groups are subtypes. The predominant group M has recognized.

Subgroups of HIV-1 [36]

Group M

- Subtype A East and Central Africa; Central Asia; Eastern Europe
- Subtype AE Southeast Asia
- Subtype AG West Africa
- Subtype B Americas; Western Europe; East Asia; Oceania
- Subtype C India; Southern and Eastern Africa
- Subtype D East Africa
- Subtype G West Africa
- Subtype H Central Africa
- Subtype J Central America
- Subtypes F, H, J, K variable

Group N Cameroon

Group O West Africa

- There are 8 distinct subtypes of HIV-2 but only A and B have become endemic. There is up to a 25% difference in genetic homology among these subtypes.

- All five subtypes can be detected by enzyme immunoassay (EIA) and Western blot assays for HIV-2 similar to those for HIV-1. [36]

MODES OF TRANSMISSION

A.   **Sexual Transmission** [38]
   ➤ Sexual, especially the heterosexual, transmission is the main driver of the epidemic in most of India, accounting for nearly 90% of nationwide prevalence.

B.   **Transmission by Blood and Blood products** [38]
   ➤ Blood transfusion is second established route of transmission after sexual route, which has reduced but is still around 2%. The people at highest risk of HIV transmission by blood to blood contact are injecting drug users (IDUs), hemophiliacs and recipients of blood transfusions. Unsafe injections, thought to contribute to about 5% of all transmissions. Transfusions or treatments with infected blood or blood products can lead to HIV transmission.

C. **Occupational Exposure** [39]
   ➤ After an exposure to blood from a patient with documented HIV infection as a result of percutaneous injury (for example, a needle stick or a cut from a sharp object), contamination of mucous membranes, or contamination of non-intact skin. (Extensive or prolonged blood contact with intact skin may constitute an occupational exposure to HIV. Among workers with percutaneous injury, the seroconversion rate after exposure to blood from source patients with AIDS was 0.44%.

D.   **Maternal-fetal transmission of HIV** [40]
   ➤ Maternal-fetal HIV-1 transmission is multifactorial, with increased risk associated both with ICD p24 antigenemia at term and with intrapartum events that increase fetal exposure to maternal blood. HIV is also transmitted from mother to infant, either intrapartum, perinatally, or via breast-feeding. [41]

REPLICATION [42]

Steps in the HIV Replication Cycle

1. Fusion of the HIV cell to the host cell surface.
2. HIV RNA, reverse transcriptase, integrase, and other viral proteins enter the host cell.
3. Viral DNA is formed by reverse transcription.
4. Viral DNA is transported across the nucleus and integrates into the host DNA.
5. New viral RNA is used as genomic RNA and to make viral proteins.
6. New viral RNA and proteins move to cell surface and a new, immature, HIV virus forms.
7. The virus matures by protease releasing individual HIV proteins.

PATHOGENESIS AND PATHOPHYSIOLOGY [36]

Persons with HIV infection can be categorized as typical progressors, rapid progressors, and non progressors toward AIDS.

A.   **Typical Progressors**: Average 8 to 10 years of "latent" HIV infection before the appearance of clinical AIDS. These persons typically have a fall in HIV viraemia following acute infection. They maintain nonsyncytium-inducing HIV variants that replicate slowly over time, until more rapidly replicating variants develop during progression to AIDS.

B.   **Rapid Progressors**: About 10% of HIV-infected persons rapidly progress to AIDS in only 2 to 3 years following initial infection. These persons have a high viral load during acute HIV infection that does not fall to the levels seen with typical progressors. They may be infected with more virulent strains of HIV. More rapid progression to AIDS is also seen in Caucasians who have the major histocompatibility complex (MHC) class I type HLA-B*35 allele

C.   **Long Term Non-Progressors**: About 10% of persons infected with HIV-1 are non-progressors, or "long survivors", who do not demonstrate a significant and progressive decline in immune function over more than 10 years. They are also termed **"elite controllers"** of HIV because they suppress viraemia below the limit of detection, even in the absence of antiretroviral therapy. They do not appear to progress to AIDS in a manner similar to the majority of HIV- infected persons. Findings in these "long survivors" include: a stable CD4 lymphocyte count, negative plasma cultures for HIV-1, fewer, HIV-infected cells, and a strong virus-inhibitory CD8+ T-lymphocyte response.

Natural history of HIV [43]

PRIMARY HIV INFECTION

There is no standard definition of primary HIV infection. However, reporting primary HIV infection, where recognized and documented, is useful and should be encouraged. Primary HIV infection can be recognized in infants, children, adolescents and adults; it can be asymptomatic or be associated with features of an acute retroviral syndrome of variable severity. Primary infection usually presents as an acute febrile illness 2–4 weeks post exposure, often with lymphadenopathy, pharyngitis, maculopapular rash, orogenital ulcers and meningoencephalitis.

Clinical Staging [43]

Clinical stage - 1

- Asymptomatic
- Persistent generalized lymphadenopathy

Clinical stage - 2

- Moderate unexplained weight loss(<10% of presumed or measured body weight)
- Recurrent respiratory tract infections sinusitis, tonsillitis, otitis media and pharyngitis)
- Herpes zoster
- Angular cheilitis
- Recurrent oral ulceration
- Papular pruritic eruptions
- Seborrhoeic dermatitis
- Fungal nail infections

Clinical stage - 3

- Unexplained severe weight loss (>10% of presumed or measured body weight)
- Unexplained chronic diarrhoea for longer than one month
- Unexplained persistent fever (above 37.6°C intermittent or constant, for longer than one month)
- Persistent oral candidiasis
- Oral hairy leukoplakia
- Pulmonary tuberculosis (current)
- Severe bacterial infections (such as pneumonia, empyema, pomposities, bone or joint infection, meningitis or bacteremia)
- Acute necrotizing ulcerative stomatitis, gingivitis or periodontitis
- Unexplained anemia (<8 g/dl), neutropenia (<0.5 × 109 per liter) or chronic thrombocytopenia (<50 × 109 per liter)

Clinical stage - 4

- HIV wasting syndrome
- Pneumocystis pneumonia
- Recurrent severe bacterial pneumonia
- Chronic herpes simplex infection (orolabial, genital or anorectal of more than one month's duration or visceral at any site)
- Oesophageal candidiasis (or candidiasis of trachea, bronchi or lungs)
- Extrapulmonary tuberculosis
- Kaposi's sarcoma
- Cytomegalovirus infection (retinitis or infection of other organs)
- Central nervous system toxoplasmosis
- HIV encephalopathy
- Extrapulmonary cryptococcosis including meningitis
- Disseminated non-tuberculous mycobacterial infection
- Progressive multifocal leukoencephalopathy
- Chronic cryptosporidiosis (with diarrhoea)
- Chronic isosporiasis
- Disseminated mycosis (coccidiomycosis or histoplasmosis)
- Recurrent non-typhoidal Salmonella bacteraemia
- Lymphoma (cerebral or B-cell non-Hodgkin) or other solid HIV-associated tumors
- Invasive cervical carcinoma
- Atypical disseminated leishmaniasis

Serological Profile of HIV [44]

- There are three recognized stages in the infection process. The first stage is the **acute phase** immediately after HIV infection. It is also referred to as the **diagnostic window** or **serological latency**. At this stage, viral nucleic acid and the HIV p24 antigen are detectable in the host serum. However, no host antibodies to HIV can be detected and the host is said to be sero negative. During the acute phase of infection, the HIV RNA levels in the blood spike at about 6 weeks post infection and then decline, while the CD4 T-cells count drops rapidly until about 6 weeks post infection when it begins a modest increase. Antibodies to HIV are usually detected in the host serum about 6–8 weeks after HIV infection, and it is unusual not to detect antibody by 3 months post infection. The detection of host antibody is referred to as sero conversion and it marks the beginning of the second stage of HIV infection known as the **chronic phase**.

- During this stage, the host antibody response to the virus evolves and matures, resulting in increasing amounts of antibody to the virus over several years.

- In addition, HIV RNA levels in the blood remain relatively stable and the host CD 4 Tcell count begins a steady decline

- The final stage of HIV infection, the onset of clinical AIDS, is distinguished by the arrival of clinical symptoms indicative of immunodeficiency, namely, contraction of opportunistic microbial infections like herpes, pneumonia and fungal infections. The inception of clinical AIDS is usually marked by a CD 4 T cell count ≤ 200 cells/mm3, which continues to decline and an increase in the number of HIV RNA copies in the blood.

Lab Diagnosis of HIV [44]

- HIV diagnostic tests function either by detecting host antibodies made against different HIV proteins or by directly detecting the whole virus itself or components of the virus (such as the HIV p24 antigen or HIV RNA).The goal of most HIV diagnostic tests is to detect HIV infection as early as possible, thereby decreasing the length of the diagnostic window. Certain HIV diagnostic tests, however, aim to distinguish between recent and longstanding HIV infections. This is done primarily for epidemiological reasons, in order to estimate the incidence of HIV.

- In addition to facilitating detection of HIV infection, serologic (blood based) tests also allow for the determination of:

  ➢ HIV viral type (HIV-1 or HIV-2)
  ➢ Viral subtype
  ➢ Viral load (as measured by viral RNA levels in the blood)
  ➢ HIV drug resistance
  ➢ How recently the viral infection was contracted (epidemiological reasons)
  ➢ Detection of anti-HIV antibodies is the mainstay of testing for HIV and diagnosis of HIV. [45]

- Tests to detect specific HIV antibodies can be classified into:
  ➢ Screening tests (ELISA/EIA and Rapid)
  ➢ Supplemental tests (ELISA/EIA and Rapid and Western Blot)

- Most common specimen collected is blood. HIV test kits are available for detecting HIV antibodies in various kinds of specimens like blood, plasma, serum, saliva and urine. Detection of HIV antibodies in specimens like saliva and urine has not been evaluated and standardized in India.

Screening Assays for HIV testing is given below [45]

ELISA (2-3 hours)

Rapid tests (minutes)

- Dot blot assays (immune concentration, vertical flow of reagents)
- Particle agglutination
- HIV spot and comb tests
- Immuno chromatography (lateral flow of reagents)
- Dipstick and comb assays (based on ELISA technology)

Other screening tests were also introduced subsequently. These include latex, red cell and gelatin particle agglutination, comb tests, line tests and dot- blot assays. These tests are easy to perform, are rapid, do not require sophisticated equipment, technical expertise and are mostly cost effective. Some of them, particularly comb tests, line tests and dot- blot assays, are also discriminatory for HIV 1 and HIV 2 antibodies.

ELISA [45]

- ELISA is the most commonly performed screening test at blood banks and tertiary care sites testing a large number of specimens a day. It is easy to perform, adaptable to large number of samples, is sensitive and specific and cost effective. There is a wide variety of ELISA assays available commercially and so the appropriate test choice can be made taking into consideration the available resource, storage facility, technical expertise available, infrastructure available, objective of testing, prevalence of infection and performance characteristics of test kits etc.

Different HIV kits (based on type of HIV antigen) available commercially [45]

- First generation of ELISA developed were very sensitive but not specific because whole viral lysates were used as antigen. These lysates usually contained small amounts of host cell components which gave rise to false positive reactions. The ELISA technologies were improved and 2nd and 3rd and 4th generation kits were developed using recombinant and synthetic peptides as antigens. So, ELISA assays available in the market may be:

- **First generation** kits use antigens derived from detergent disruption of viruses grown in human lymphocytes.

- **Second generation** kits use artificially derived recombinant antigens expressed from bacteria or fungi.

- **Third generation** kits use chemically synthesized oligopeptides of about 15-40 amino acids (synthetic peptides).

- **Fourth generation** kits use a combination of recombinant and synthetic peptides and can detect both HIV antigen (p24) and antibodies concurrently.

Principles of ELISA [45]

On the basis of the principle of the test ELISA can be divided into:
- Indirect

- Competitive

- Sandwich and

- Capture assays.

All ELISAs consist of either HIV antigen or antibody (depending upon the principle) attached on a solid phase (matrix or support) and, incorporate a conjugate and substrate detection system. Viral antigens may be whole virus lysates, recombinant or synthetic peptides. The matrix can be "wells" or "strips" of a microplate, plastic beads or nitrocellulose paper. Conjugates are most often antibodies (IgG, sometimes IgM and IgA also) coupled to enzymes (alkaline phosphatase or horseradish peroxidase), fluorochromes or other reagents that will subsequently bring about a reaction that can be visualised. In case of enzyme conjugates the signal generated is a colour reaction and in case of fluorochromes it is fluorescence.

Indirect ELISA [45]

- HIV antigens are attached covalently to the solid phase support allowing HIV antibodies present in the specimen to bind, and these bound antibodies are subsequently detected by enzyme labeled anti-human immunoglobulin and specific substrate system. If the test specimen contained antibodies colour reaction will take place. Indirect ELISA is the most commonly used system

Competitive ELISA [45]

- In this assay the HIV-antibodies present in the specimen compete with the enzyme conjugated antibodies in t h e reagent for binding to the antigen on the solid phase. If the test specimen contains HIV antibodies, these will compete with the labeled antibodies in the reagent for binding to antigen. So that less or not labeled antibodies can attach to the solid phase. Hence, faint or no colour is produced on addition of substrate if specimen contained HIV antibodies.

Sandwich ELISA [45]

- This is a modification of indirect ELISA to improve sensitivity and specificity of the test. Antigen bound to the solid phase binds antibody in the test specimen in first step. Since antibody molecules are bivalent they are still able to bind to another molecule. The next step is addition of similar enzyme labeled HIV antigen i.e. same antigen as on solid phase. This will attach to the antibody molecule which is already bound to the solid phase antigen with one arm. Thus forms a sandwich of antigen + antibody + enzyme labeled antigen complex. The next step is addition of specific substrate which results in development of colour which is measured by ELISA reader. One big advantage of sandwich ELISA is that all classes of HIV-antibodies can be detected.

Antigen and antibody capture ELISA [45]

- Antigen capture ELISA can be based on principle of indirect or competitive ELISA, only difference being in the initial step of attaching antigen to the solid phase in case of indirect ELISA.

- A monoclonal antibody directed against an HIV antigen is bound to the solid support. Next step is addition of HIV antigen supplied as reagent. This antigen is captured by the monoclonal antibody bound to the solid phase. Test specimen appropriately diluted is added next. HIV antibodies if present in the specimen bind to HIV antigen on solid support. Remaining principle is same as indirect ELISA. Only advantage of antigen capture ELISA is that it is more specific than indirect assay.

False positive and False negative ELISA results [45]

- There are some conditions other than HIV infection which may give a reactive HIV results i.e. false positive result. Also, sometimes the result may be negative even in HIV infected i.e. false negative result. Some of the common conditions giving erroneous results are listed below.

False positive result:

- Auto-immune diseases
- Multiple pregnancies
- Multiple transfusions
- Antibody to Class II HLA-Ag (HLA-DR4)
- Hyper gammaglobulinemia
- Antipolystyrene antibodies
- Chronic alcoholics
- Patients with hepatitis
- Hepatitis B immunization
- Technical error etc.
- Others

False negative result:

- Infected but not yet seroconverted, window period
- Late stage disease (immune collapse)
- Technical error

Rapid assays [45]

- A number of rapid assays based on principle of agglutination, immunoconcentration (vertical flow), immunochromatography (lateral flow) and ELISA based have been developed for ease of performance and quick results. These assays generally require less than 30 minutes to perform and do not require special equipments. These tests have been validated in India. The results of the blinded multicenteric study showed the rapid tests to be as sensitive and specific as the ELISA in the field. Consequently the tests were sanctioned for use at ICTCs, PPTCTCs and emergency situations as the screening as well as supplementary/

confirmatory tests following the national strategies/algorithms of testing.

Agglutination assays [45]

- Agglutination assays incorporate a variety of antigen coated carriers like red cells, latex particles, gelatin particles and microbeads. These particles are used to support or carry the antigen. HIV antigens are attached to the carrier particles by non- specific attachment.

- Agglutination assays have good sensitivity, do not require sophisticated equipment, are easy to perform, require no wash procedures and are cost effective. However, specificity is somewhat compromised and prozone reaction may be seen. To overcome the prozone reaction, diluted specimen is used to perform the test. During the agglutination reaction HIV antibody combines with HIV antigen on the carrier particles and since all antibodies are multivalent, a sort of lattice network is formed which can be visualised macroscopically or microscopically as per the directions of the manufacturer.

Supplemental (confirmatory) assays [45]

- Supplemental/confirmatory test are undertaken to confirm the HIV infection status of an individual who is either symptomatic or asymptomatic and has history of high risk behavior.

- Supplemental assays are used for the purpose of diagnosis of HIV infection and identification of the individual i.e. the individual is informed about the test results in the proper way (with counseling and confidentiality).

- Various supplemental assays available commercially are listed below E/R with different antigen system (recombinant or synthetic peptides) or with different principle of test which makes the test more specific

  ➢ Western blot (WB)

➢Immuno blot
➢Line immunoassay
➢Indirect fluorescent antibody test (IFA)
➢Radio immune precipitation test (RIPA)

Western blot / Immunoblot / Line immunoassay [45]

- •These are most widely accepted supplemental assays, are highly specific but are expensive, labor intensive, need expertise to interpret and may also give unequivocal / indeterminate results. The specificity of these tests is based on two factors: separation of antigens and their concentration.

- •IB and WB use viral antigens from whole virus lysates electrophoretically transferred to a membrane support. So these blots may contain contaminating cellular components.

- •Recombinant or synthetic HIV antigens mechanically applied onto the support membrane are used in Line Immunoassays (LIA). These do not contain contaminating cellular components and are highly specific.

- •Results are interpreted either as per WHO criteria i.e. presence of at least two envelope bands (gp 120, gp160, gp 41 etc.) or as per the manufacturer's criteria for positive, indeterminate and negative.

Strategies/Algorithms of HIV testing in India [45]

- •Because of the enormous risk involved in transmission of HIV through blood, safety of blood and blood products is of paramount importance. Since the PPV is low in populations with low HIV prevalence, WHO / GOI have evolved strategies / algorithms to detect HIV infection in different population groups and to fulfill different objectives.

- •   S
trategy/algorithm I: Serum is subjected to one E/R for HIV. If negative, the serum is considered free of HIV and if positive, the sample is taken as

HIV infected for all practical purposes. This strategy is used for ensuring donation safety (blood/blood products organ, tissues, sperms etc.). The unit of blood testing reactive (positive) is discarded. Donor is informed only if written consent for knowing the result was taken. Donor is given the provisional positive report and is referred to ICTC/VCTC for counseling/confirmation of test result. In India this strategy is followed. Others are Strategy/algorithm II A, Strategy/algorithm II B, Strategy/algorithm III respectively as per WHO.

HEPATITS B

- Hepatitis B is a potentially life-threatening liver infection caused by the hepatitis B virus. It is a major global health problem and the most serious type of viral hepatitis. It can cause chronic liver disease and puts people at high risk of death from cirrhosis of the liver and liver cancer.

HISTORICAL ASPECT OF HEPATITIS B VIRUS

- Hepatitis epidemics, which likely included hepatitis B as one cause, have spanned the course of human history, dating back to antiquity and observations on an epidemic of jaundice by Hippocrates. Yet it wasn't until 1883 that a German scientist first described what was later thought to be this particular form of viral hepatitis in a group of people who had developed jaundice after receiving a smallpox vaccine prepared from human blood. Similarly, an outbreak of hepatitis-related jaundice affecting approximately 50,000 U.S. Army personnel during World War II was later attributed to HBV infection transmitted through a contaminated yellow-fever vaccine, based on research performed in the late 1980s. The infectious agent responsible for these hepatitis outbreaks, the hepatitis B virus, was identified by Dr. Baruch Blumberg while working at the NIH in the 1960s—a discovery that later earned him the Nobel Prize in Physiology or Medicine in 1976.

- Basic research on the natural history of HBV infection led to the preparation in the 1970s and 1980s of the first hepatitis B vaccines based on heat-inactivated and blood plasma-derived viruses.

EPIDEMIOLOGY

- It is estimated two billion people have been infected with the hepatitis B virus and more than 240 million have chronic (long-term) liver infections. About 600,000 people die every year due to the acute or chronic consequences of hepatitis B. [46]

- Approximately 1 in 12 persons worldwide, or some 500 million people, are living with chronic viral hepatitis; 1 million of those who are infected die each year, primarily from cirrhosis or liver cancer resulting from their hepatitis B and hepatitis C infections. Many of those who are chronically infected with viral hepatitis are unaware of their infection as the virus can go 20 to 30 years before they develop symptoms or feel sick. [47]

Hepatitis B Virus [48]

- Hepatitis B virus is an enveloped DNA virus belonging to the family Hepadna viridae. It is unrelated to any other human virus; however, related hepatotropic agents have been identified in woodchucks, ground squirrels, and kangaroos. The complete virion is a 42-nm, spherical particle that consists of an envelope around a 27-nm core.

- The core comprises a nucleocapsid that contains the DNA genome.

- The viral genome consists of partially double-stranded DNA with a short, single stranded piece. It comprises 3200 nucleotides, making it the smallest DNA virus known. Closely associated with the viral DNA is a DNA polymerase. Other components of the core are a hepatitis B core antigen (HBcAg) and the hepatitis B *e* antigen (HBeAg), which is a low molecular weight glycoprotein. The 42-nm particle is the "Dane particle" or the hepatitis B virus. The 22-nm particles are the filamentous and circular forms of hepatitis B surface antigen (HBsAg) or protein coat.

• The envelope of the virus contains the hepatitis B surface antigen (HBsAg), which is composed of one major and two other proteins. Antigenically there exist a group-specific determinant, termed *a,* and a number of subtypes that are important in epidemiologic typing, but not in immunity, because there is antigenic cross-reactivity and cross-protection between subtypes. Aggregates of HBsAg are often found in great abundance in serum during infection. They may assume spherical or filamentous shapes with a mean diameter of 22 nm and may contain portions of the nucleocapsid.

GENOTYPES [49]

•   H BV genotypes represent naturally occurring strains of HBV that have evolved over the years and reflect the geographical distribution of HBV throughout the world. Up to now, eight different HBV genotypes have been identified and shown to cluster in different areas of the world. They display an 8% inter-group divergence in the complete nucleotide sequence of HBV and differences in the nucleotide homology of the surface gene, which result in different hepatitis B surface antigen (HBsAg) serotypes.

• Genotype A is mainly found in Northwestern Europe, North America and Africa, whereas genotypes B and C have been described in South-Eastern Asian populations. Genotype E and F are seen in East Africa and the New World, respectively. Genotype D is most often found in southern Europe, parts of Central Asia, India, Africa and the Middle East. Genotype G is a recently determined genotype in France, America, and Germany while genotype H has been reported in patients from Central America

REPLICATION CYCLE [48]

•The replication of hepatitis B virus involves a reverse transcription step, and, as such, is unique among DNA viruses. The double-stranded DNA is organized as two strands. One, a short strand, is associated with the viral DNA polymerase and is of positive polarity. The complete or long strand is complementary and thus of negative polarity. In viral replication, full-length positive viral RNA transcripts are inserted into maturing core particles late in the replicative cycle. These mRNA strands form a template for a reverse transcription step in which negatively stranded DNA is synthesized. The RNA template strands are then degraded by ribonuclease activity. A positive-stranded DNA is then synthesized, although this is not completed prior to virus maturation and release and thus results in the variable-length short positive DNA strands found in the virion.

PATHOGENESIS [48]

•In the past, hepatitis B was known as post-transfusion hepatitis or as hepatitis associated with the use of illicit parenteral drugs (serum hepatitis). However, over the past few years it has become clear that the major mode of acquisition is through close personal contact with body fluids of infected individuals. HBsAg has been found in most body fluids, including saliva, semen, and cervical secretions. Under experimental conditions, as little as 0.0001 mL of infectious blood has produced infection. Transmission is therefore possible by vehicles such as inadequately sterilized hypodermic needles or instruments used in tattooing and ear piercing. The factors determining the different clinical manifestations of acute hepatitis B are largely unknown; however, some appear to involve immunologic responses of the host. The serum sickness–like rash and arthritis that may precede the development of symptoms and jaundice appear to be related to circulating immune complexes that activate the complement system. Antibody to HBsAg is protective and associated with resolution of the disease. Cellular immunity also may be important in the host response, because patients with depressed T-lymphocyte function have a high frequency of chronic infection with the hepatitis B virus. Antibody to the HBcAg, which appears during infection, is present in chronic carriers with persistent hepatitis B virion production and does not appear to be protective.

•The virus has not been shown to possess a transforming gene but may well activate a cellular oncogene. It is also possible that the virus does not play such a direct molecular role in oncogenicity, because the natural history of chronic hepatitis B infection involves cycles of damage or death of liver cells interspersed with periods of intense regenerative hyperplasia. This significantly increases the opportunity for spontaneous mutational changes that may activate cellular oncogenes.

MANIFESTATIONS

•The clinical picture of hepatitis B is highly variable. The incubation period may be as brief as 7 days or as long as 160 days (mean, approximately 10 weeks). Acute hepatitis B is usually manifested by the gradual onset of fatigue, loss of appetite, nausea and pain, and fullness in the right upper abdominal quadrant. Early in the course of disease, pain and swelling of the joints and occasional frank arthritis may occur. Some patients develop a rash. With increasing involvement of the liver, there is increasing cholestasis and, hence, clay-colored stools, darkening of the urine, and jaundice. Symptoms may persist for several months before finally resolving.

MALARIA

- Malaria is caused by a small living organism, called a parasite, which infects a person's red blood cells. It is transmitted from one person to another by the bite of female Anopheles mosquitoes. The parasite must go through a complex cycle in both the mosquito and in humans before transmission can take place. [51]

Epidemiology

- Malaria imposes great socio-economic burden on humanity and with six other diseases like diarrhoea, HIV/AIDS, tuberculosis, measles, hepatitis B and pneumonia account for 85% of Global infectious disease burden. About 36% of the world population, i.e. 2020 million is exposed to the risk of contracting malaria in ~ 90 countries. World Health Organization estimates 300–500 million malaria cases annually and 90% of this burden is in Africa alone. [52]

- Even a century after the discovery of malaria transmission through mosquitoes in India by Sir Ronald Ross in 1897, malaria continues to be one of India's leading public health problems. Each year approximately 2.5 million cases and 4,000 deaths are reported but the disease is grossly underestimated due to cases being seen by the private sector and otherwise not being included in the malaria control program reporting system. [52]

- In India, nine Anopheline vectors are involved in transmitting malaria in diverse geo-ecological paradigms. About 2 million confirmed malaria cases and 1,000 deaths are reported annually, although 15 million cases and 20,000 deaths are estimated by WHO South East Asia Regional Office. India contributes 77% of the total malaria in Southeast Asia. Multi-organ involvement/dysfunction is reported in both Plasmodium falciparum and P. vivax cases. Most of the malaria burden is borne by economically productive ages. [53]

- Transfusion-transmitted malaria occurs at an estimated rate of 0.25 cases per 1 million blood units collected.

- T

  he malaria parasite life cycle involves two hosts. During a blood meal, a malaria-infected female Anopheles mosquito inoculates sporozoites into the human host. Sporozoites infect liver cells and mature into schizont, which rupture and release merozoites. (Of note, in Plasmodium vivax and P. ovale, a dormant stage [hypnozoites] can persist in the liver and cause relapses by invading the bloodstream weeks, or even years later.) After this initial replication in the liver (exo-erythrocytic schizogony), the parasites undergo asexual multiplication in the erythrocytes (erythrocytic schizogony). Merozoites infect red blood cells. The ring-stage trophozoites mature into schizonts, which rupture, releasing merozoites. Some parasites differentiate into sexual erythrocytic stages (gametocytes). Blood-stage parasites are responsible for the clinical manifestations of the disease.

- T

  he gametocytes, male (micro gametocytes) and female (macro gametocyte) are ingested by an Anopheles mosquito during a blood meal. Multiplication in the mosquito is known as the sporogonic cycle. While in the mosquito's stomach, the microgametes penetrate the macrogametes, generating zygotes. The zygotes in turn become motile and elongated (ookinetes) and invade the midgut wall of the mosquito where they develop into oocysts. The oocysts grow, rupture, and release sporozoites, which make their way to the mosquito's salivary glands. Inoculation of the sporozoites into a new human host perpetuates the malaria life cycle.

Pathogenesis [55]

- Plasmodium falciparum causes more severe disease than the other Plasmodium species do. Several features of P. falciparum account for its greater pathogenicity:

- P. falciparum is able to infect red blood cells of any age, leading to high parasite burdens and profound anemia. The other species infect only young or old red cells, which are a smaller fraction of the red cell pool.

- P. falciparum causes infected red cells to clump together (rosette) and to stick to endothelial cells lining small blood vessels (sequestration), which blocks blood flow. Several proteins, including P. falciparum erythrocyte membrane protein 1 (PfEMP1), form knobs on the surface of red cells PfEMP1 binds to ligands on endothelial cells.

- Ischemia due to poor perfusion causes the manifestations of cerebral malaria, which is the main cause of death due to malaria in children.

- P. falciparum stimulates production of high levels of cytokines, including TNF, IFN-γ, and IL-1. GPI-linked proteins, including merozoite surface antigens, are released from infected red cells and induce cytokine production by host cells by a mechanism that is not yet understood.

- These cytokines suppress production of red blood cells, increase fever, stimulate nitric oxide production (leading to tissue damage), and induce expression of endothelial receptors for PfEMP1 (increasing sequestration).

Host Resistance to Plasmodium [55]

- T
here are two general mechanisms of host resistance to Plasmodium.

- First, inherited alterations in red cells make people resistant to Plasmodium.

- Second, repeated or prolonged exposure to Plasmodium species stimulates an immune response that reduces the severity of the illness caused by malaria.

Clinical Presentation [56]

- Malaria is characterized by recurrent paroxysms of chills and high fever. They begin with chills and sometimes headache, followed by a high, spiking fever with tachycardia, often accompanied by nausea, vomiting and abdominal pain. The high fever produces marked vasodilation and often associated orthostatic hypotension. The patient defervesces after several hours and is usually exhausted and drenched in sweat.

Other modes of Malaria Transmission

- This history of transfusion malaria dates from 1884 when Gerhardt demonstrated on two human subjects that malaria can be transmitted by blood inoculation. [57]

- **Blood transfusion (Transfusion malaria):** This is fairly common in endemic areas. Following an attack of malaria, the donor may remain infective for years (1-3 years in P. falciparum, 3-4 years in P. vivax, and 15-50 years in P. malariae.) Most infections occur in cases of transfusion of blood stored for less than 5 days and it is rare in transfusions of blood stored for more than 2 weeks. Frozen plasma is not known to transmit malaria.

- **Mother to the growing fetus (Congenital malaria):** Intrauterine transmission of infection from mother to child is well documented. Placenta becomes heavily infested with the parasites. Congenital malaria is more common in first pregnancy, among non - immune populations.

- **Needle stick injury:** Accidental transmission can occur among drug addicts who share syringes and needles.

Diagnosis - Microscopy

- Microscopy of stained thick and thin blood smears remains the gold standard for confirmation of diagnosis of malaria. The advantages of microscopy are:

  - ➤ The sensitivity is high.
  - ➤ It is possible to detect malaria parasites at low densities.
  - ➤ It also helps to quantify the parasite load.
  - ➤ It is possible to distinguish the various species of malaria parasite and their different stages.

- In malaria microscopy, two kinds of blood film are used: thick and thin. [51]

- **The thick film** - Always used to search for malaria parasites. The film consists of many layers of red and white blood cells. During staining, the haemoglobin in the red cells dissolves (dehaemoglobinization), so that large amounts of blood can be examined quickly and easily. Malaria parasites, when present, are more concentrated than in a thin film and are easier to see and identify.

- **The thin film** - It is used to confirm the malaria parasite species, when this cannot be done in the thick film. It is used to search for parasites only in exceptional situations. A well-prepared thin film consists of a single layer of red and white blood cells spread over less than half the slide. The frosted end of the slide is used for labeling.

RAPID DIAGNOSTIC TESTS (RDTs) [58]

• These tests are based on the detection of antigens derived from malaria parasites in lysed blood, using immunochromatographic methods. Most frequently they employ a dipstick or test strip bearing monoclonal antibodies directed against the target parasite antigens. The tests can be performed in about 15 minutes. Several commercial test kits are currently available.

• Other diagnostic methods are available, but they are not as suitable for wide field application as microscopy or RDTs and are unsuitable for use in routine disease management. They include microscopy using fluorochromes; polymerase chain reaction (PCR) based tests and antibody detection by serology.

• **Microscopy using fluorochromes** such as acridine orange, either on blood smears or on centrifuged blood specimens (**QBC®** technique) is expensive and requires special equipment and supplies (centrifuge and centrifuge tubes, special light sources and filters) **PCR** is more sensitive and specific than all other techniques. It is, however, a lengthy procedure that requires specialized and costly equipment and reagents, as well as laboratory conditions that are often not available in the field. **Antibody detection by serology** only measures prior exposure and not specifically current infection.

HEPATITIS C

Hepatitis C Virus [60]

- It is impossible to really know the origins of hepatitis C since there are no stored blood samples to test for the virus that are older than 50 years. However, given the nature of the evolution of all viruses, hepatitis C has probably been around for hundreds of thousands of years or more before evolving into the current strains. Some experts speculate that since HGV/GBV-C, a close relative of HCV, originated in Old and New World primates, the beginnings of HCV might be traced back to 35 million years ago. However, this is just speculation and it is impossible to corroborate these theories at the present time.

- In the 1980's, investigators from the Centers for Disease Control (headed up by Daniel W. Bradley) and Chiron (Michael Houghton) identified the virus. In 1990, blood banks began screening blood donors for hepatitis C.

Epidemiology of HCV [61]

- Although HCV is endemic worldwide, there is a large degree of geographic variability in its distribution. Countries with the highest reported prevalence rates are located in Africa and Asia; areas with lower prevalence include the industrialised nations in North America, northern and Western Europe, and Australia. Populous nations in the developed world with relatively low rates of HCV seroprevalence include Germany (0·6%), Canada (0·8%), France (1·1%), and Australia (1·1%). Low, but slightly higher seroprevalence rates have been reported in the USA (1·8%), Japan (1·5–2·3%), and Italy (2·2%).

- The epidemiology of hepatitis C in India has not been studied systematically. Most of the studies of the prevalence of hepatitis C have been based in blood banks with the assumption that the blood donors are a surrogate for the population at large. [62]

•HCV infection in India has a population prevalence of around 1%, and occurs predominantly through transfusion and the use of unsterile glass syringes. HCV genotypes 3 and 2 are prevalent in 60-80% of the population and they respond well to a combination of interferon and ribavirin. About 10%-15% of CLD and HCC are associated with HCV infection in India. HCV infection is also a major cause of post- transfusion hepatitis. [63]

Virology of HCV [64]

•HCV is a member of the Flaviviridae family. It is a small, enveloped, single-stranded RNA virus with a 9.6-kilobase (kb) genome that codes for a single polyprotein with one open reading frame, which is subsequently processed into functional proteins. The 5′ end of the genome encodes a highly conserved nucleocapsid core protein, followed by envelope proteins E1 and E2. Two hypervariable regions (HVR 1 and 2) are present in the E2 sequence. A protein, $p7$, is believed to function as an ion channel. Toward the 3′ end are six less conserved nonstructural proteins: NS2, NS3, NS4A, NS4B, NS5A, and NS5B. NS5B is the viral RNA-dependent RNA polymerase. The 3′ sequences of both the positive- and negative-strand RNAs contribute cis-acting functions that are essential for viral replication. The secondary structure and protein-binding properties of these highly conserved non translated regions are thought to promote HCV RNA synthesis and genome stability through the binding of various host and viral proteins.

•Because of the poor fidelity of the HCV RNA polymerase (NS5B), the virus is inherently unstable, giving rise to multiple genotypes and subtypes. Indeed, within any given patient HCV circulates as a population of divergent but closely related variants known as quasispecies. Over time, dozens of quasispecies can be detected within one individual and mapped as derivative strains of the original HCV strain that infected the patient. In particular, elevated titers of anti-HCV IgG occurring after an active infection do not consistently confer effective immunity. A characteristic feature of HCV infection, therefore, is repeated bouts of hepatic damage, the result of reactivation of a preexisting infection or emergence of an endogenous, newly mutated strain.

Genotype [65]

- Hepatitis C is divided into six distinct genotypes throughout the world with multiple subtypes in each genotype class. A genotype is a classification of a virus based on the genetic material in the RNA (Ribonucleic acid) strands of the virus. Generally, patients are only infected with one genotype, but each genotype is actually a mixture of closely-related viruses called quasi-species. These quasi-species have the ability to mutate very quickly and become immune to current treatments, which explains why chronic Hepatitis C is so difficult to treat.

- Following is a list of the different genotypes of chronic Hepatitis C:

  - Genotype 1a
  - Genotype 1b
  - Genotype 2a, 2b, 2c & 2d
  - Genotype 3a, 3b, 3c, 3d, 3e & 3f
  - Genotype 4a, 4b, 4c, 4d, 4e, 4f, 4g, 4h, 4i & 4j
  - Genotype 5a
  - Genotype 6a

Genotype patterns

- It is believed that the hepatitis C virus has evolved over a period of several thousand years. This would explain the current general global patterns of genotypes and subtypes:

  - 1a - mostly found in North & South America; also common in Australia
  - 1b - mostly found in Europe and Asia.
  - 2a - is the most common genotype 2 in Japan and China.
  - 2b - is the most common genotype 2 in the U.S. and Northern Europe.
  - 2c - the most common genotype 2 in Western and Southern Europe.
  - 3a - highly prevalent here in Australia (40% of cases) and South Asia.
  - 4a - highly prevalent in Egypt
  - 4c - highly prevalent in Central Africa
  - 5a - highly prevalent only in South Africa

➢6a - restricted to Hong Kong, Macau and Vietnam
➢7a and 7b - common in Thailand
➢8a, 8b & 9a - prevalent in Vietnam
➢10a & 11a - found in Indonesia

HCV RNA replication [66]

• The process of HCV RNA replication is poorly understood. The key enzyme for viral RNA replication is NS5B, an RNA-dependent RNA polymerase (RdRp) of HCV. After the RdRp has bound to its template, the NS3 helicase is assumed to unwind putative secondary structures of the template RNA in order to facilitate the synthesis of minus-strand RNA. In turn, the newly synthesized antisense RNA molecule serves as the template for the synthesis of numerous plus-stranded RNA. The resulting sense RNA may be used subsequently as genomic RNA for HCV progeny as well as for polyprotein translation. Another important viral factor for the formation of the replication complex appears to be NS4B, which is able to induce an ER derived membranous web containing most of the non-structural HCV proteins including NS5B.

Modes of Transmission [67]

• Transfusion of blood and blood products from non-tested blood donors
• Organ transplantation from infected donors
•   IV drug use; therapy with injected drugs with contaminated (or not safe)
• equipment
•   Hemodialysis; occupational exposure to blood
•   Perinatal infection
•   Sexual transmission
•   Moreover, due to the great variety of human activities with potential exposure to blood, several possible biologic transmission models exist, such as tattoo, piercing, barber shop, scarification rituals, circumcision, and acupuncture.

• **Transfusion of blood products**: Transfusion of blood and blood products from non- tested donors is considered the most important type of transmission. However, after randomization of pre-donation screening processes, a significant reduction in HCV transmission through blood products transfusion is observed. It has been estimated that, between 1960 and 1991, 5% to 15% of blood product receptors were infected with HCV and that, currently, after adoption of screening tests, the risk of infection from blood transfusion is around 0.001% per unit of blood transfused.

• **Intravenous drug use**: After reduction in HCV transmission by blood products transfusion, sharing contaminated material by IV drug users became the greatest risk factor for transmission of disease. Intravenous drug use was one of the main types of HCV transmission in the last 40 years in countries like the United States and Australia [12, 20], being currently the main risk factor in developed countries. [12, 20] In these countries, IV drug use is responsible for approximately 70% to 80% of HCV contaminations in the last 30 years.

• **Medical procedures and nosocomial exposure**: Injectable therapies with contaminated (or unsafe) equipment represent another possible form of HCV transmission. Despite the scarcity of reliable data, it has been estimated that approximately two million individuals are infected annually by this route. In developing countries, the supply of sterilized material can be inadequate or nonexis- tent. Moreover, outside of medical centers, injectable therapies might be performed by untrained individuals; therefore, throughout life, a person can received several injections with contaminated material, increasing significantly the accumulate risk of HCV infection. Patients on hemodialysis have higher prevalence of HCV infection, ranging from 19% to 47.2%.

• **Solid organ transplantation**: The estimated prevalence of HCV infection in organ transplant recipients is complicated by the influence of immunosuppression on the accuracy of serological tests commonly used. The prevalence of anti-HCV in organ donors, according to studies in cadavers, ranges from 4.2% to 5.1%.

- **Occupational exposure:** Needle sticks accidents with percutaneous inoculation is a well-documented HCV transmission, with seroconversion rates after a single percutaneous exposure to known infected material ranging from 3% to 10%.

- **Vertical transmission:** Rates of vertical HCV transmission range from 0% to 20%, with a mean of approximately 5% in most studies. Risk factors for vertical transmission include elevated maternal viral load, prolonged labor, internal fetal monitoring, and HIV-HCV coinfection. Coinfected mothers were 3.8 times more prone to transmit HCV to the fetus. Breast feeding did not contribute significantly to HCV transmission.

- **Sexual transmission:** The risk associated with sexually transmitted HCV is not yet fully understood, and this risk factor is one of the most controversial in the epidemiology of hepatitis C among different results in different studies.

Pathogenesis of HCV [68]

- HCV infection is characterized by its propensity to evolve into chronicity and by a wide clinical spectrum. About 85% of patients infected by HCV will develop chronic infection and resolution of acute hepatitis C is observed in only 15%. The severity of the liver disease varies widely from asymptomatic chronic infection, with normal liver tests and nearly normal liver, to severe chronic hepatitis, leading rapidly to cirrhosis and hepatocellular carcinoma.

- The quality of the cellular immune response is crucial for the elimination or the persistence of HCV infection. CD4+ T cells and their cytokines with inflammatory and regulatory activities seem to play an important role in the immunopathogenesis of chronic HCV infection. CD4+ T cell responses are polarized into type 1 and type 2 helper T cell (Th1 and Th2) responses. The Th1 cells secrete interleukin 2 (IL-2) and interferon gamma, which are important stimuli for the development of the host antiviral immune responses, including cytotoxic T- lymphocyte (CTL) generation and NK-cell activation. The Th2 cells produce IL-4 and IL-lo, which enhance antibody production and down regulate the Th1 response. It is hypothesized that the imbalance between the Th1 and the Th2 responses is implicated in disease progression and the inability to clear infection.

Clinical manifestations and natural history of HCV infection [69]

- The spectrum of clinical manifestations of HCV infection varies in acute versus chronic disease. Acute infection with HCV is most often asymptomatic. It leads to chronic infection in about 80% of cases. The manifestations of chronic HCV range from an asymptomatic state to cirrhosis, and hepatocellular carcinoma. HCV infection usually is slowly progressive. Thus, it may not result in clinically apparent liver disease in many patients if the infection is acquired later in life. Approximately 20.30% of chronically infected individuals develop cirrhosis over a 20-30 year period of time.

HIV-HCV Co-infection in India [62]

- Both HIV and HCV infection share the same routes of transmission and it is not surprising that co-infection of these viruses is common. The prevalence of hepatitis C infection in patients with HIV infection has been very variable. Two studies from Lucknow and Chennai showed relatively low rates of co-infection of 1.61% and 2.2% respectively (Saravanan et al 2007; Tripathi et al 2007). Both these studies were done in patients with low incidence of IV drug use.

Laboratory Testing [70]

- Two classes of assays are used in the diagnosis and management of HCV infection:

  - Serologic assays that detect specific antibody to hepatitis C virus (anti-HCV)
  - Molecular assays that detect viral nucleic acid. These assays have no role in the assessment of disease severity or prognosis.

Serological Assays

- Tests that detect anti-HCV are used both to screen for and to diagnose HCV infection. Anti-HCV can be detected in the serum or plasma using a number of immunoassays. The specificity of current EIAs for anti-HCV is greater than 99%. [70] With 2nd generation enzyme-linked immunoassays (EIAs), HCV specific antibodies can be detected approximately 10 weeks after infection. To narrow the diagnostic window from viral transmission to positive serological results, a 3rd generation EIA has been introduced that includes an antigen from the NS5 region and/or the substitution of a highly immunogenic NS3 epitope, allowing the detection of anti- HCV antibodies approximately four to six weeks after infection with a sensitivity of more than 99% . [71]

- Following the discovery of HCV and the sequencing of its genome in 1989, the **first generation of HCV ELISAs** was produced using recombinant proteins complementary to the NS4 region of the HCV genome as antigens. These assays showed limited sensitivity and specificity.

- **Second generation tests**, which included recombinant or synthetic antigens from the putative core and nonstructural regions NS3 and NS4 resulted in a marked improvement in sensitivity and specificity.

- **The third generation tests** include antigens from the NS5 region of the genome, in addition to those used in second generation assays. Third generation tests have improved sensitivity, though this has been shown to be more likely due to the improvements to the core and NS3 antigens rather than the inclusion of the NS5 antigen. However, despite these improvements, the time between infection with HCV and the appearance of detectable antibodies (window period) is generally more than 40 days It is anticipated that test kits will undergo further improvement in the future.

- **Quantitative HCV Core Antigen Assay** [71] - This assay comprises 5 different antibodies, is highly specific (99.8%) and shows somewhat less sensitivity for determination of chronic hepatitis C as HCV RNA measurement.

- **<u>Nucleic Acid Testing for HCV</u>** [71] - Because of the importance of an exact HCV RNA load determination for therapeutic management, the World Health Organization (WHO) established the HCV RNA international standard based on international units (IU) which is used in all clinically applied HCV RNA tests. Currently, several HCV RNA assays are commercially available.

- **Qualitative HCV RNA tests** [71] - Include the qualitative RT-PCR, is an FDA- and CE approved RT-PCR system for qualitative HCV RNA testing that allows detection of HCV RNA concentrations down to 50 IU/ml of all HCV genotypes.

- **Transcription-mediated amplification- (TMA)** [71] - based qualitative HCV RNA detection has a very high sensitivity (lower limit of detection 5-10 IU/ml). A commercially available TMA assay is the Versant™ HCV RNA Qualitative Assay (Siemens, Germany).

- **HCV RNA quantification** can be achieved either by target amplification techniques (competitive and real-time PCR) or by **signal amplification techniques** (branched DNA (bDNA) assay). [71]

SYPHILIS

- The venereal form of treponematosis, caused by the spirochete Treponema pallidum, plagued every major city in the preantibiotic era. "Civilization means syphilization," was an idea touted by Richard von Krafft-Ebing in the late 19th, and early 20th centuries that the effects of modern life make men more susceptible to syphilis and other diseases. Christopher Columbus was thought of as an importer of syphilis to Europe. [72]

- Syphilis was even mentioned in Act 3 of Timon of Athens by William Shakespeare. The name for syphilis is derived from Fracastorius' 1530 epic poem in three parts, Syphilis sive morbus gallicus ("Syphilis or The French Disease"), about a shepherd boy named Syphilus who insulted the sun god of Haiti and was punished by that god with a horrible disease.

- Treponema pallidum was demonstrated under magnification in Berlin on March 3, 1905 by Schauddin and Hoffman. The discovery had come at the right time because syphilis was epidemic and presented a very serious health problem in all industrialized countries at that time. During the early 1900s, syphilis was estimated to affect ~10% of the population of the United States and Western Europe. [73]

Columbian hypothesis of origin of syphilis [73]

- At least three hypotheses about origin of syphilis in Europe were advanced. Most popular and best supported by historical, archeological (skeletal remains), and molecular phylogenetic studies is so-called Columbian hypothesis, which suggests that original treponemal disease spread from Africa through Asia, entering North America. Approximately 8 millennia later, it mutated to syphilis. Ample presence of the skeletal evidence of syphilis at the site of the Columbus' landing (Dominican Republic) suggests that the Columbus soldiers got infected there and then transmitted the disease to the Old

World when they returned in 1462.

Syphilis in India

- Syphilis, known in India as Portuguese disease or firanga or firangi roga reached the subcontinent in early 16th century and soon became widespread. A comparison of the present data with that seen a decade back (1986 - 1990) in the same region revealed a sharp decline from 14. 6% to 9.26% in syphilis. [75]

Morphology of Treponema [73]

- Treponema pallidum belongs to the only order Spirochaetales in class of bacteria Spirochaetes. All spirochetes are helical or spiral shaped microorganisms. Spirochetes are highly motile and propel themselves in corkscrew manner by rotating around their longitudinal axes. These organisms can swim easily through gel-like materials that hinder most other flagellated organisms, which allow some spirochetes to occupy unique ecological niches, such as sediments in pond and lake bottoms, the guts of certain arthropods, and the rumens of cows and sheep. The arrangement of spirochete flagella is also unique among the bacteria. The flagella are inserted subterminally at each end of the cell, wrap around the protoplasmic cylinder, and usually overlap in the center region. They are located between the protoplasmic cylinder and an outer membrane-like structure called the outer sheath and called endoflagella or periplasmic flagella.

Challenges for research [73]

- Treponema pallidum cannot be cultivated long term (more than 100-fold ~7 generations) in vitro. In laboratory, T. pallidum can be only maintained by propagation in rabbits. Because of fragility of its outer membrane, researches are unable to modify the bacterium genetically in order to conduct experiments, which in many other bacteria clarified various aspects of their biology such as protein functions, mechanisms of virulence, and others.

Infectivity [73]

- Treponema pallidum is transmitted by direct contact, usually sexual. Infection is initiated when T. pallidum penetrates dermal micro abrasions or intact mucous membranes. Studies have shown that 16 to 30% of individuals who have had sexual contact with a syphilis-infected person become infected. Actual transmission rates can be higher. The 50% infectious dose is estimated to be only 57 organisms. Upon initial infection the parasites prefer to multiply at the point of entry causing the inflammatory response and formation of characteristic chancre. From the chancre the treponemes disseminate rapidly to the blood and lymphatic's and make their way to different parts of the body including Central Nervous System (CNS). T. pallidum has been shown to induce the production of matrix metalloproteinase-1 (MMP-1) in dermal cells. MMP-1 is involved in breaking down collagen, which may help T. pallidum to traverse the junctions between endothelial cells and penetrate tissues.

Clinical Features [76]

- The clinical course of syphilis is divided into 4 stages - primary, secondary, and tertiary stages in which characteristic manifestations occur and a latent stage in which the patient is asymptomatic but seropositive. The latent stage occurs with the resolution of the secondary stage. Identifying the appropriate stage of disease is important because it affects duration of treatment.

Primary syphilis

- Primary syphilis is acquired by direct contact with open lesions of an individual with syphilis. T. pallidum is transmitted through small abrasions in the skin. The initial manifestation of primary syphilis is called a chancre, a painless skin ulceration at the point of exposure, often on the penis, vagina, or rectum and usually ranging in size from 0.3 to 3.0 cm. The average time from exposure to the appearance of the chancre is 3 weeks. The chancre may persist for 4 to 6 weeks and usually heals spontaneously even without treatment. Inguinal lymphadenopathy can occur.

Secondary Syphilis

- Secondary syphilis is the result of hematogenous dissemination of T. pallidum, which subsequently causes systemic symptoms. This stage typically occurs 6 to 8 weeks after the primary infection. Manifestations of secondary syphilis can mimic many other diseases, and cutaneous manifestations are common. Manifestations include a generalized rash, fever, malaise, myalgias, alopecia, and generalized lymphadenopathy. The initial rash of secondary syphilis begins as a faint exanthema with macular pink lesions that are asymptomatic and are often overlooked. Therash is not pruritic, and lesions are oval macules of varying size (0.5–1 cm) a few days, a symmetric papular eruption can appear. In 50% to 80% of cases, the papular rash of secondary syphilis involves the palms of the hands and the soles of the feet.

Latent Syphilis

- Because transmission usually occurs during the primary and secondary stages when lesions are present, it is important to diagnose and treat syphilis early. Latent syphilis is defined as having serologic evidence of infection without signs or symptoms of disease. Latent syphilis can be classified as early or late depending on the time frame from initial infection (early, < 1 year and late, ≥ 1 year). The distinction is important because it affects treatment duration.

Tertiary Syphilis

- Tertiary (late) syphilis generally occurs 5 to 20 years after initial infection and can present with mucocutaneous, cardiac, ophthalmologic, neurologic, or osseous abnormalities.4 Tertiary disease will develop in approximately one third of untreated individuals, manifesting as gummatous (benign) syphilis, cardiovascular syphilis, or neurosyphilis.

Neurosyphilis

- Neurosyphilis can occur at almost any stage of syphilis. Classic syndromes of neurosyphilis are now much less common in the antibiotic era, likely due to inadvertent antibiotic treatment of early stages of syphilis while treating other infections. Neurologic findings are varied in neurosyphilis. Early neurosyphilis can present with meningeal and meningovascular involvement within weeks to years of primary infection. Meningovascular neurosyphilis can compromise the vascular supply of the central nervous system, resulting in infarctions. Meningeal syphilis can present with syphilitic meningitis and manifest with headache, nausea, vomiting, and neck stiffness in the absence of fever. Seizures can also occur.

Laboratory Diagnosis of Syphilis [77]

Methods for the direct detection of syphilis include

- Rabbit infectivity testing (RIT)
- Dark field (DF) or microscopy following immunostaining (direct fluorescent antibody/silver staining)
- Most recently polymerase chain reaction (PCR).

- RIT can be used for blood, cerebrospinal fluid (CSF,) amniotic fluid, primary and secondary lesion exudate, and lymph node (LN) aspirate. The technique is very sensitive and very specific, capable of detecting as few as 1 to 2 organisms. However, it is only available in a limited number of research settings and, although in these settings it has proven to be valuable, it is not practical as a diagnostic tool.

- **Dark Field microscopy** detects Treponema pallidum (Tp) based upon characteristic morphology and motility. It can be used for primary and secondary lesions (except oral lesions), exudate, LN aspirate, CSF, amniotic fluid, and other fluids. DF microscopy is a very valuable tool as it is sensitive, inexpensive, and can be performed at the point of care.

- The sensitivity of Dark Field microscopy depends on the state of lesion development, but can reasonably be expected to detect approximately 105 Tp/mL2. Specificity is highly dependent on the skill of the microscopist, making training and the maintenance of quality assurance programs very important.

- **DFA-TP (direct fluorescent antibody staining for Treponema pallidum)**, is an immunofluorescence enzyme-based microscopy method that can be used for lesions smears, concentrated fluids, tissue brushings, and fixed or unfixed tissues. The specificity of the technique depends on the type of primary antibody used (polyclonal, monospecific, monoclonal). Sensitivity depends upon the concentration of Treponema pallidum in the sample. Monoclonal antibodies are better for touch preps while polyclonal antibodies are optimal for formalin-fixed tissues.

- **PCR** identifies Treponema pallidum by amplifying organism-specific DNA or RNA sequences. PCR can be performed on lesion swabs, Lymph Node aspirates, CSF, blood, amniotic fluid, and fixed or unfixed tissue samples. However, the sensitivity varies greatly by specimen type. PCR can theoretically detect 1 gene copy and the specificity, while intrinsically very high, depends on primer selection, skill of the laboratorian, and sample type, quality, and handling

Serologic Testing for the Diagnosis of Syphilis in Adults

- Syphilis serologic diagnosis relies on testing for non treponemal and treponemal antibodies. These antibodies differ markedly with respect to antigenic reactivities and kinetics during the disease process.

VDRL, RPR, USR and TRUST tests [78]

- The VDRL and USR (Unheated Serum Reagin) tests are microflocculation tests and are read under a microscope.

- A disadvantage of the VDRL test is that the antigen suspension must be prepared fresh daily, whereas the USR test uses a stabilized antigen.

Testing algorithm for primary syphilis

- VDRL test is the only nontreponemal test that can be used to test CSF due to the limited sensitivity and specificity of the other nontreponemal tests.

- The RPR and TRUST (Toluidine Red Unheated Serum Test) tests are macroscopic flocculation tests and require no microscope.

- The RPR test uses a stabilized suspension of VDRL antigen to which charcoal particles are added to aid in the visualization of the test reaction. The RPR test is one of the most commonly used non treponemal tests, and is a simplified version of the VDRL test.

- In TRUST test, particles of toluidine red are used in place of the charcoal particles of the RPR test.

- Each of the above tests can be used as a quantitative test.

REFERENCES

1. Jasani J. Sero-prevalence of transfusion transmissible infections among blood donors in a tertiary care hospital. 2012.
2. Bihl F, Castelli D, Marincola F, Dodd RY, Brander C. Transfusion-transmitted infections. J Transl Med. 2007; 6(5): 25.
3. İmdat Dilek CD, Ali Bay, Hayrettin Akdeniz, Ahmet Faik Öner. Seropositivity rates of HBsAg, anti-HCV, anti-HIV and VDRL in blood donors in Eastern Turkey. Turk Journal of Haemat 2007; 24 (1- March): 4-7.
4. Lt Col PK Gupta CHK, Mr DR Basannar, Brig M Jaiprakash. Transfusion Transmitted Infections in Armed Forces: Prevalence and Trends. MJAFI. 2006; 62: 348-50.
5. Gagandeep Kaur SB, Ravneet Kaur , Paramjit Kaur , Shailja Garg. Patterns of infections among blood donors in a tertiary care centre: A retrospective study. The Nat Med J of India, 2010. p. 147-9.
6. Chandra T, Kumar A, Gupta A. Prevalence of transfusion transmitted infections in blood donors: an Indian experience. Tropical doctor. 2009; 39(3): 152-4.
7. Chattoraj A, Behl R, Kataria V. Infectious disease markers in blood donors. Med J Armed Forces India. 2008; 64: 33-5.
8. Sawke N, Sawke G, Chawla S. Seroprevalence of common transfusion-transmitted infections among blood donors. 2010.
9. Wenz B, Mercuriali F, Aubuchon JP. Practical methods to improve transfusion safety by using novel blood unit and patient identification systems. American journal of clinical pathology. 1997; 107(4 Suppl 1): S12.
10. Greenwalt T. A short history of transfusion medicine. Transfusion. 1997; 37(5): 550-63.
11. Myhre BA. James Blundell - pioneer transfusionist. Transfusion. 1995; 35(1): 74-8.
12. Denis JB. A Letter Concerning a New Way of Curing Sundry Diseases by Transfusion of Blood: Written to Monsieur de Montmor, Counsellor to the French King, and Master of Requests, 1967.
13. Lower R. The success of the experiment of transfusing the blood of one animal into another. Philos Trans R Soc Lond. 1666(1): 352.

14.Denis J. An extract of a letter: Touching a late cure of an inveterate phrenisy by the transfusion of blood. Philos Trans R Soc Lond. 1668(3): 617-23.

15.Denis J. An extract of a printed letter.... touching the differences risen about the transfusion of blood. Philos Trans R Soc Lond. 3: 710-5.

16.Jones H, Mackmul G. The influence of James Blundell on the development of blood transfusion. Annals of Medical History. 1928; 10: 242-8.

17.Blundell J. Experiments on the transfusion of blood by the syringe. Medico- chirurgical transactions. 1818; 9(Pt 1): 56.

18.Oberman H. Early history of blood substitutes: Transfusion of milk. Transfusion. 1969; 9(2): 74-7.

19.Lewisohn R. A new and greatly simplified method of blood transfusion. Med Rec. 1915; 87: 141.

20.Lewisohn R. A new and greatly simplified method of blood transfusion.Med Rec 1915; 87: 141-2.

21.Robertson OH. Transfusion with preserved red blood cells. British Medical Journal. 1918; 1(2999): 691-5.

22.Telischi M. Evolution of Cook County Hospital Blood Bank. Transfusion. 1974; 14(6): 623-8.

23.Loutit J, Mollison P. Disodium-Citrate—Glucose Mixture as a Blood Preservative. British Medical Journal. 1943; 2(4327): 744-5.

24.Gibson JG, Gregory CB, Button LN. Citrate-phosphate-dextrose solution for preservation of human blood: a further report. Transfusion 1961; 1: 280–287.

25.Sally V. Rudmann. Textbook of blood banking and transfusion medicine. Philedelphia: WB Saunders; 1995.

26.Marfatia Y, Sharma A, Modi M. Overview of HIV/AIDS in India. Indian Journal of Sexually Transmitted Diseases and AIDS. 2007; 28(1): 1.

27.Global HIV and AIDS estimates, 2009 and 2010. UNAIDS (2010) 'Unite for universal access: Overview brochure on 2011 High Level Meeting on AIDS'2011 [updated 2010]; Available from: http://www.avert.org/worldstats.htm.

28.NACO. Annual Report New Delhi, 2010-2011.

29.Welfare MoHaF. Prevalence of HIV/AIDS. New Delhi; Available from: http://pib.nic.in/newsite/PrintRelease.aspx?relid=67292.

30.Karnataka GO. Annual Action Plan 2011-2012. In: Society KSAP, editor. Karnataka, 2012, p. 13-21.

31.Goura Kudesia TW. Clinical and Diagnostiv Virology Cambridge University Press; 2009.

32.Jawetz M, Adelberg. AIDS and Lentiviruses. In: Brooks GF, editor. Medical Microbiology. 25th edition, LANGE; 2010.

33.Rubin R. Rubin's pathology: Clinicopathologic foundations of medicine. Lippincott Williams & Wilkins; 2012, p. 156.

34.Taylor BS, Sobieszczyk ME, McCutchan FE, Hammer SM. The challenge of HIV-1 subtype diversity. New England Journal of Medicine. 2008; 358(15): 1590-602.

35.Klatt EC. Pathology of AIDS. 2nd edition, Florida State University College of Medicine; 2009.

36.NACO. Annual Report New Delhi 2010-2011. Available from: http://www.nacoonline.org.

37.Sharma R. Profile of HIV-positives and determinants with mode of transmission of HIV/AIDS patients on anti-retroviral treatment center at civil hospital, Ahmedabad. Indian journal of sexually transmitted diseases. 2011; 32(1): 14.

38.Tokars JI, Marcus R, Culver DH, Schable CA, McKibben PS, Bandea CI, et al. Surveillance of HIV infection and zidovudine use among health care workers after occupational exposure to HIV-infected blood. Annals of internal medicine. 1993; 118(12): 913-9.

39.Boyer PJ, Dillon M, Navaie M, Deveikis A, Keller M, O'Rourke S, et al. Factors predictive of maternal-fetal transmission of HIV-1. JAMA: the journal of the American Medical Association. 1994; 271(24): 1925-30.

40.Linden JV, Bianco C. Blood safety and surveillance: Informa HealthCare; 2001.

41.HIV Replication Cycle. Available from: http://www.niaid.nih.gov/ topics/ HIVAIDS/ Understanding/Biology/pages/hivreplicati oncycle.aspx.

42.Organization WH. WHO case definitions of HIV for surveillance and revised clinical staging and immunological classification of HIV-related disease in adults and children. Geneva: World Health Organization. 2006.

43.Iweala OI. HIV diagnostic tests: an overview. Contraception. 2004; 70(2): 141- 7.

44.Guidelines for HIV Testing. New Delhi: National AIDS Control Organisation; 2008 [cited 2007, 29 august]; Available from: http://nacoonline.org/upload/Final%20Publications/Blood%20Safety/GUI LDELINES %20FOR%20HIV%20TESTING.pdf.

45.Heptatitis B Fact Sheet [database on the Internet]. WHO. 2011. Available from: http://www.who.int/mediacentre/factsheets/fs204/en/.

46.CDC. World Hepatitis Day – July 28th. Atlanta, USA Center's for Disease Control and Prevention; 2011; Available from: http://www.cdc.gov/features/dshepatitisawareness/index.html#References.

47.Sherris JC. Hepatites B Virus. In: RYAN KJ, editor. Medical microbiology: an introduction to infectious diseases. 4th ed: Elsevier Biomedical Press BV; 2004. p. 544-49.

48.Abdo AA, Al-Jarallah BM, Sanai FM, Hersi AS, Al-Swat K, Azzam NA, et al. Hepatitis B genotypes: relation to clinical outcome in patients with chronic hepatitis B in Saudi Arabia. World J Gastroenterol. 2006; 12(43): 7019-24.

49.Pathology Q. A Guide to Hepatitis Serology. [cited 2009 March]; Symbion Pathology Pvt. Ltd]. Available from: http://www.qml.com.au/Files/Hep_Ser_485.pdf.

50.WHO. BASIC MALARIA MICROSCOPY, 2010. Available from: http://whqlibdoc.who.int/publications/2010/9789241547918_eng.pdf.

51.Dash A. Estimation of true malaria burden in India. A Profile of National Institute of Malaria Research 2nd edition, New Delhi, India: National Institute of Malaria Research. 2009: 91-9.

52.Kumar A, Valecha N, Jain T, Dash AP. Burden of malaria in India: retrospective and prospective view. The American journal of tropical medicine and hygiene. 2007; 77(6 Suppl): 69-78.

53.Hillyer CD, Silberstein LE, Ness PM, Anderson KC, Roback JD. Blood banking and transfusion medicine. Recherche. 2007; 67: 02.

54.Robbins SL, Kumar V, Abbas AK, Cotran RS, Fausto N. Robbins and Cotran pathologic basis of disease. WB Saunders Company; 2010. p. 386-88.

55.Rubin R. Rubin's pathology: Clinico-pathologic foundations of medicine. Lippincott Williams & Wilkins; 2012. p. 404-05.

56.Bruce-Chwatt LJ. Transfusion malaria revisited. Tropical diseases bulletin. 1982; 79(10): 827.

57.Organization WH. New perspectives malaria diagnosis. The Organization; 2001, p. 12-13.

58.Organization WH. New perspectives malaria diagnosis. The Organization; 2001, p. 14-15.

59. HCV Education & Support: A Brief History of Hepatitis C2010; (VERSION 5.1): Available from: http://www.hcvadvocate.org/ hepatitis/ factsheets_pdf/Brief_History_HCV_10.pdf.

60. Shepard CW, Finelli L, Alter MJ. Global epidemiology of hepatitis C virus infection. The Lancet infectious diseases. 2005; 5(9): 558-67.

61. Mukhopadhya A. Hepatitis C in India. Journal of biosciences. 2008; 33(4): 465- 73.

62. Acharya S, Madan K, Dattagupta S, Panda S. Viral hepatitis in India. The National medical journal of India. 2006; 19(4): 203.

63. Robbins SL, Kumar V, Abbas AK, Cotran RS, Fausto N. Robbins and Cotran pathologic basis of disease. 8th edition, WB Saunders Company; 2010. p. 847-48.

64. Available from: http://www.hepatitis-central.com/hepatitis-c/hepatitis-c-genotypes.html.

65. Thomas Berg CB. HCV Structure and Viral Replication. In: Mauss B, Rockstroh, Sarrazin, Wedemeyer, editor. Short Guide to Hepatitis C: Flying Publisher; 2012, p. 24-6.

66. Tatiana Martins JLN-S, Leonardo de Lucca Schiavon. Epidemiology of hepatitis C virus infection. Rev Assoc Med Bras. [Review]. 2011; 57(1): 105-10.

67. Boyer N, Marcellin P. Pathogenesis, diagnosis and management of hepatitis C. Journal of hepatology. 2000; 32: 98-112.

68. Mauss S. Hepatology: a clinical textbook: Flying Publisher; 2010.

69. Ghany MG, Strader DB, Thomas DL, Seeff LB. Diagnosis, management, and treatment of hepatitis C: an update. Hepatology. 2009; 49(4): 1335-74.

70. Sarrazin CLaC. Diagnostic Tests in Acute and Chronic Hepatitis C. In: Mauss Rockstroh, Sarrazin, Wedemeyer, editor. Short Guide to Hepatitis C: Flying Publisher; 2012. p. 28-33.

71. Sehgal VN, Verma P, Chatterjee K, Chaudhuri A, Chatterjee G, Rasool F. Origin and Evolution of Syphilis: Drifting Myth. WITH GENTLE EFFICACY1. 2010; 10: 8-12.

72. Treponema pallidum. Available from: http://www.metapathogen.com/ syphilis/.

73. Organization WH. Global prevalence and incidence of selected curable sexually transmitted infections: overview and estimates. World Health Organization; 2002, p. 19-25.

74. Thappa D, Kaimal S. Sexually transmitted infections in India: Current status (except human immunodeficiency virus/ acquired immunodeficiency syndrome). Indian Journal of Dermatology. 2007; 52(2): 78.

75. Karnath BM. Manifestations of syphilis. Hosp Phys. 2009: 43-8.

76. Laboratory Diagnostic Testing for Treponema pallidum Atlanta, GA: Expert Consultation Meeting Summary Report2009 January 13-15, 2009.

77. Ratnam S. The laboratory diagnosis of syphilis. The Canadian Journal of Infectious Diseases & Medical Microbiology. 2005; 16(1): 45.

78. Voluntary Blood Donation Programme- An Operational Guideline. New Delhi National AIDS Control Organisation; 2007.

79. Giri PA, Deshpande JD, Phalke DB, Karle LB. Seroprevalence of transfusion transmissible infections among voluntary blood donors at a tertiary care teaching hospital in rural area of India. Journal of Family Medicine and Primary Care. 2012; 1(1): 48.

80. Bahadur S, Jain S, Goel RK, Pahuja S, Jain M. Analysis of blood donor deferral characteristics in Delhi, India. Southeast Asian Journal of Tropical Medicine and Public Health. 2009; 40(5).

81. Patel PA, Patel SP, Oza H. Seroprevalence of transfusion transmitted infections (ttis) in blood donors at western ahmedabad–a secondary care hospital based study. 2012.

82. Deshpande RH, Bhosale S, Gadgil P, Sonawane M. Blood Donor s Status of HIV, HBV, HCV and Syphilis in this Region of Marathwada, India. Journal of Krishna Institute of Medical Sciences University.1.

83. Kulkarni N. Analysis of the seroprevalence of HIV, HBsAg, HCV and syphilitic infections detected in the pretranfusion blood: A short report. International Journal of Blood Transfusion and Immunohematology (IJBTI). 2012; 2: 1-3.

84. Amrutha Kumari B, Deepa S, Venkatesha D. Blood Transfusions: Are They Life Saving or Transfusing Infections. Online J Health Allied Scs. 2011; 10(2): 7.

85. Makroo R, Chowdhry M, Bhatia A, Arora B, Rosamma N. Prevalence of HIV among blood donors in a tertiary care centre of north India. The Indian Journal of Medical Research. 2011; 134(6): 950.

86. Hepatitis B. WHO; 2012 [cited 2012 July]; Available from: http://www.who.int/mediacentre/factsheets/fs204/en/.

87. Pahuja S, Sharma M, Baitha B, Jain M. Prevalence and trends of markers of hepatitis C virus, hepatitis B virus and human immunodeficiency virus in Delhi blood donors: A hospital based study. Japanese journal of infectious diseases. 2007; 60(6): 389.

88. Soldan K, Davison K, Dow B. SUPPLY IN THE UNITED KINGDOM, 1996 TO 2003. Euro Surveill. 2005; 10(2): 17-9.

89. Afsar I, Gungor S, Sener A, Yurtsever S. The prevalence of HBV, HCV and HIV infections among blood donors in Izmir, Turkey. Indian Journal of Medical Microbiology. 2008; 26(3): 288.

90. Fasola FA, Kotila TR, Akinyemi JO. Trends in transfusion-transmitted viral infections from 2001 to 2006 in Ibadan, Nigeria. Intervirology. 2008; 51(6): 427-31.

91. Singh K, Bhat S, Shastry S. Trend in seroprevalence of Hepatitis B virus infection among blood donors of coastal Karnataka, India. The Journal of Infection in Developing Countries. 2009; 3(05): 376-9.

92. Short Guide to Hepatitis C In: Mauss B, Rockstroh, Sarrazin, Wedemeyer, editor. 2012 edition, The Flying Publisher; 2012, p. 15.

93. Thakral B, Marwaha N, Chawla Y, Saluja K, Sharma A, Sharma R, et al. Prevalence & significance of hepatitis C virus (HCV) seropositivity in blood donors. The Indian Journal of Medical Research. 2006; 124(4): 431.

94. Jain A, Rana S, Chakravarty P, Gupta R, Murthy N, Nath M, et al. The prevalence of hepatitis C virus antibodies among the voluntary blood donors of New Delhi, India. European journal of epidemiology. 2003; 18(7): 695-8.

95. Sgaier S, Mony P, Jayakumar S, McLaughlin C, Arora P, Kumar R, et al. Prevalence and correlates of Herpes Simplex Virus-2 and syphilis infections in the general population in India. Sexually transmitted infections. 2011; 87(2): 94-100.

96. Ampofo W, Nii-Trebi N, Ansah J, Abe K, Naito H, Aidoo S, et al. Prevalence of blood-borne infectious diseases in blood donors in Ghana. Journal of clinical microbiology. 2002; 40(9): 3523-5.

97. Matee MIN, Magesa PM, Lyamuya EF. Seroprevalence of human immunodeficiency virus, hepatitis B and C viruses and syphilis infections among blood donors at the Muhimbili National Hospital in Dar Es Salaam, Tanzania. BMC Public Health. 2006; 6(1): 21.

98.Tessema B, Yismaw G, Kassu A, Amsalu A, Mulu A, Emmrich F, et al. Seroprevalence of HIV, HBV, HCV and syphilis infections among blood donors at Gondar University Teaching Hospital, Northwest Ethiopia: declining trends over a period of five years. BMC Infectious diseases. 2010; 10(1): 111.

99.Stanier P, Taylor DL, Kitchen AD, Wales N, Tryhorn Y, Tyms AS. Persistence of cytomegalovirus in mononuclear cells in peripheral blood from blood donors. BMJ: British Medical Journal. 1989; 299(6704): 897.

100.Dubey A, Elhence P, Ghoshal U, Verma A. Seroprevalence of malaria in blood donors and multi-transfused patients in Northern India: Relevance to prevention of transfusion-transmissible malaria. Asian Journal of Transfusion Science. 2012; 6(2): 174.

101.Choudhury N, Ramesh V, Saraswat S, Naik S. Effectiveness of mandatory transmissible diseases screening in Indian blood donors. The Indian Journal of Medical Research. 1995; 101: 229.

102.Contreras CE, Pance A, Marcano N, González N, Bianco N. Detection of specific antibodies to Plasmodium falciparum in blood bank donors from malaria- endemic and non-endemic areas of Venezuela. The American journal of tropical medicine and hygiene. 1999; 60(6): 948-53.

103.Srikrishna A, Sitalakshmi S, Damodar P. How safe are our safe donors? Indian Journal of Pathology and Microbiology. 1999; 42(4): 411.

104.Bahadur S, Pujani M, Jain M. Use of rapid detection tests to prevent transfusion-transmitted malaria in India. Asian Journal of Transfusion Science. 2010; 4(2): 140.

105.Uneke CJ, Ogbu O, Nwojiji V. Potential risk of induced malaria by blood transfusion in South-eastern Nigeria. McGill Journal of Medicine: MJM. 2006; 9(1): 8.

www.ingramcontent.com/pod-product-compliance
Lightning Source LLC
Chambersburg PA
CBHW021901170526
45157CB00005B/1916